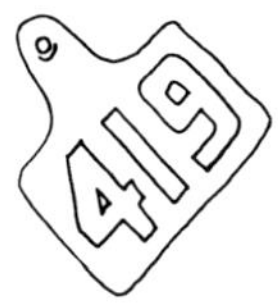

Scream

scre

eam

or never minding

Lia Purpura

The Literary House Press
Washington College
Chestertown, Maryland

Design by Jehanne Dubrow and Lindsay Lusby

ISBN: 978-0-937692-24-0

The Literary House Press is an imprint
of the Rose O'Neill Literary House
at Washington College
in Chestertown, Maryland

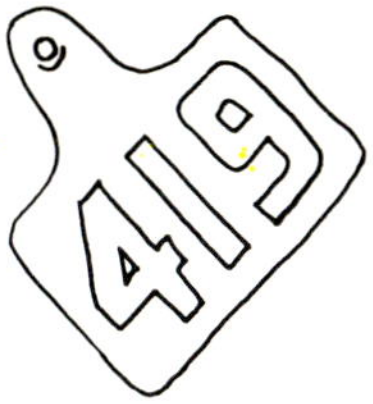

THERE ARE THINGS I'M SUPPOSED TO NEVER MIND. "Never mind" means *silent* and *agreed upon*, and that I must want, more than anything, to get through the day, and so should assent to go along. Glance. Turn the page. Turn away from a scream, and the place from which a scream would rise, if cultivated by attention paid.

Subjects one might avoid: ruined land, ruined animals. Because the issues of the day can begin to feel old, and people get tired of feeling bad.

When I was a child I was not daunted. I let myself get completely exhausted.

Never minding makes it possible to do things like eat what you want, and talk about simple, daily things.

A scream is not speech.

Edvard Munch's *The Scream* was recently sold for nearly 120 million dollars. He called it "a cry from the heart," and wrote about it to a friend that "I was being stretched to the limit...Nature was screaming in my blood. I was at breaking point..."[1] But as a gesture performed over and over, on coffee mugs, tote bags, key chains, and cards, it's much reduced, quieted so as to be understood. Seeing the scream again and again, we agree not to.

Instead, we refer to.

Consider all we throw away. The tin my mints came in could do so much work. Could be put to good use and serve again, holding buttons, coins, pills. Then fewer tins would have to be made. Imagine (though there's no need to, this is all real) how many things are

1. From Sue Prideaux's *Edvard Munch: Behind the Scream*, Yale University Press, 2005.

made to be thrown away. You can't care about them. Their brevity isn't meaningful like, say, a dart with a poisoned tip, a spear, an arrowhead—objects whose single use sustains.

Yes, I understand making tins is a job. A way of making a living. That people have jobs making trash for a living.

That subs or heroes or grinders, whatever they're called in a given place, are sold by the inch. That drinks are called "bottomless." That for a set price you can eat all night, stack BBQ ribs ten inches high if you balance just right, walking back to your seat.

What a deal that is.

If more is the measure, if the point is *a lot*—best not to fuss over the origins of stuff. And, too, if origins are questionable, you'd want the distance between farm and table to be as vast as possible. Vast is stable. Ribs are tasty. I mean a *factory* farm. A Concentrated Animal Feeding Operation. A CAFO. An acronym is a form of speed, a way to fly past the origin of an idea.

"Concentrate" was my favorite cereal as a kid. I liked it, too, for its double meaning—a dense substance/a command to think hard. Here, though, *Concentrated* means: twenty hogs in a space the size of your bedroom; ten chickens in a two-foot square pen—that's an area the size of this page for each chicken. Under such conditions, animals are driven mad.

I probably don't have to be so direct here.

I'm sure you've heard a lot about this.

Once, a bandwagon was used for exactly that purpose: to carry musicians in a procession. To *jump on one* now means *to join a successful enterprise*. So many forms of success depend on never minding, on taking the steps two at a time, up to the wagon and climbing on for the ride. Or think of riding a tide: a force absorbs you, purpose transports, and a shared mind washes over.

At the edges, though—near jetties and inlets, in dips and depressions—little tide pools settle and still, and that's where the interesting stuff collects. You have to get down on your knees to see all the briny, colloidal, fast-swimming creatures that, at a distance, look only like murk.

I want to think about #419.

What might seem like veering around here isn't. I'm trying to lay in how many times a day, and in how many ways, a person—I—might turn away.

Or else, what—stay and scream?

Solastalgia is a very good word, made by combining the Latin *solacium* (comfort) and the Greek root *-algia* (pain). Philosopher Glenn Albrecht created it to define "the pain experienced when the place one loves and where one resides is under assault." I'm working on a word for *the loss of fellow-feeling for a creature and the strange emptiness that loss leaves in its place.* "Zoosympenia" might do. "Zoo" (animal) + "sym" (feeling) + "penia" (loss). But first, before words, a feeling must root. For me, it was winter, late afternoon, when eye level with the stove—a beautiful old Chambers—I'd set things in motion. The oven door had a hooked lever which, fixed between two orange dials, made it into the face of an owl. It was possible to come around the corner unseen, inch up to the bird, flip its beak, hear it talk, touch its ever-open eyes. Say *hi* in passing, or play in its gaze. Nearby was a horse in the linoleum. A horse's *head* to be more precise—a little disconcerting but, like a ragged cloud shape in the sky, more a suggestion than a truncated thing. The horse had its own scent when I laid my head next to it. We talked in the mornings. It was always ready. I also had a collection of bees, paper ones my mother cut from tuna cans and distributed on alternate weeks to me and my sister. We'd toss them in air and watch them sift down on the flowers we were.

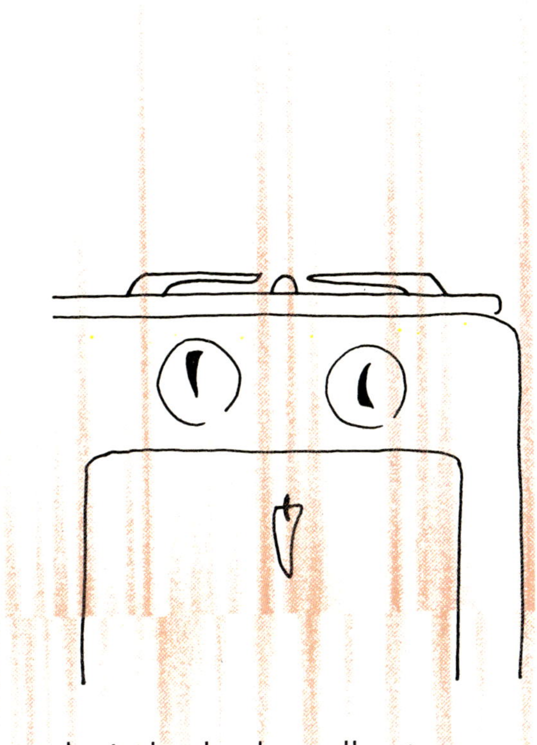

When I was four, the world was ending. I wasn't certain. I just suspected. It was 1968. The war was on. I had a dog, a monkey, a fox. I did not think of them as toys. If the world was a storm, I was an ark.

I'll get to work on another word, too—something for *the loss of relationship to singular objects due to an overabundance of them.*

How about: *Aesthesioplegia.* "Aesthesio" (sensation) + "plegia" (paralysis).

To understand an object's habits, its tricks, you have to live with it daily. A milk bottle (the one I grew up with had a high, cinched waist and a full-skirted body) might let you take an illegal swig, but only if you used both hands. You might blue the last inch of milk by holding it up to the morning sun and tipping it almost horizontal.

Endless abundance clobbers the chance for relationships. So for example, if asked *Would you like to help stop wrecking the earth*, you'd say, of course, *Yes*. If asked, then, to drink only from fountains and never again buy a bottle of water...well, it's hard to imagine giving up convenience—though what you'd get, in return, is the chance to learn the quirks of your local water fountains: the cold ones with high arcs, the calmer but warmer, dribbly ones. You might choose to walk a bit further for the beautiful, pebbly fountain, or make do with closer, slightly tin-flavored water. Or you might carry your own collapsible cup (I had one as a Girl Scout), which when folded would be exactly the size of a tin of mints.

#419 is a cow; that's a tag in its ear. There's a #308 right behind it, a #376, and a #454—all jammed in the frame of the photo. This must be a mixed lot. If I stand back just a little or, rather, hold the newspaper out at arm's length and unfocus a bit, the numbers fade and the cows are wearing bell-shaped earrings. If I shut my eyes, and shut many more things—doors in the brain, as if windows in cold—if I conjure up Heidi and green fields and milk pails, I can hear the little cowbells tinkling. And see the concrete outside my door roll up like a rug, cartoon style, revealing the sleeping pasture below. Which reconstitutes in sun and springs with fat flowers back to life. And feeds all the cows on endless, rollicking, hand-colored cartoon greens.

A *Starry Night* mug. A Caillebotte trivet. *Mona Lisa* fridge magnets. Cézanne wrapping paper.

419 is bending to eat. Not grass but corn. I guess I should say,

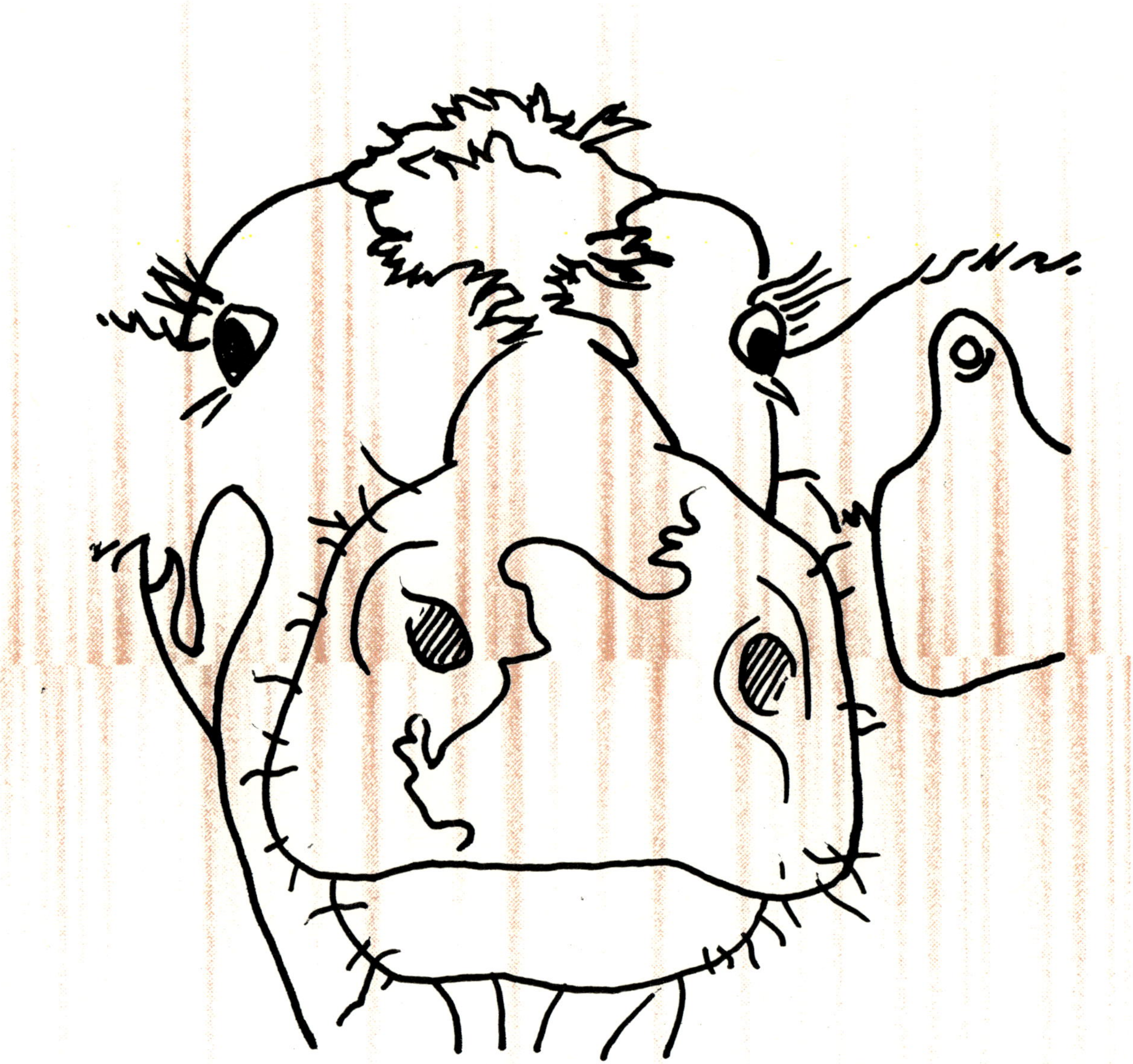

she's eating what she's made to eat, which likely includes swine manure, poultry feathers, cement dust, and plastic for roughage. All cheaper and more efficient than grass, easily supplied in confined spaces—though contributing, as you might imagine, to conditions that justify prophylactic antibiotics.

But let me just do a description here. Establish a scene. 419 is bending to eat. Her tags are clearly visible, as is the patch of white between her eyes, which are good at seeing blue-purples and yellow-greens and not so adept at the red-brown-pink spectrum. And there, in relief, against the white chest of 308—the fringe of 419's black eyelashes. Like a girl's. Like my son's in sleep.

Once you really see a thing (even briefly, and slight as a lash) it's hard to unsee.

This summer I saw two sandpipers dig for crabs at the edge of the shore with long, pointed beaks. When the waves rolled out, one dug, picked, and ate very quickly. The other plucked its catch and ran straight to a dry spot higher up and took its time pecking every bit of meat. One found more crabs, but ate each only partly. The other stuck with a single crab and mined it completely. One was not more adept, or the other wiser—just that there was, undeniably, the *inclination* each had. It was impossible to miss: each preferred a particular method, each had an idea, a disposition. A sensibility. An imagination.

Endless *Starry Night*s wreck time. *Great Wave*s get drained, and *Sunflowers* dimmed by repetition. What Munch once made of a sensation at dusk (his friends having left him after a walk, his brief pause on a path and his sense of despair) is no longer a space where you and the painter might linger together. Now it's a trinket. A T-shirt. A necklace. A thing you stop seeing that stands in for. It's a joke. A tactic. A way to connect at the office, in meetings (which make everyone want to scream). Why tame that original moment? Why encourage skipping so quickly past? Why reduce so much to mild laughter? A while back, inflatable *Scream* dolls were all the rage—there on your desk or life-size, a bald, angsty friend who'd commiserate, lift a beer with you at the end of the day. Imagine him on the label of a microbrew—the colors amped, the border crisp. A beer called Scream. Scream lager. Scream ale.

I'd like to think the painting's new owner wanted back to a

time before the image refracted, to park a chair near the original sky, the fat blue scream pressing in like a wave, the bridge's red rail like an iron poker, the gash of sky marbled and fleshy, and to be, as everything was—clouds, air, nerves, smudgy suggestion of boats—overcome. I'm hoping he wanted to clear some space, amid colors eddying fast and draining, so a scream might speak, singularly, to him.

To own a thing in a more perfect way, go into training. Adopt the gestures of a beloved friend you rarely see, the way she holds a thought in her hands and twists it into place in air. Study how her fingers flare when describing something unjust, or a point beyond which she won't be pushed, so you, too, can shape ideas in air and close up the distance between you.

Or, paring a spot of mold from cheese, take up a worn, well-sharpened knife, and cutting slowly toward your body, let the blade come to rest against the flesh of your thumb—and in this, the old-fashioned way, free-hand, with no cutting board—spend a few moments with your grandmother again.

Some forms of ownership don't require the purchase of anything.

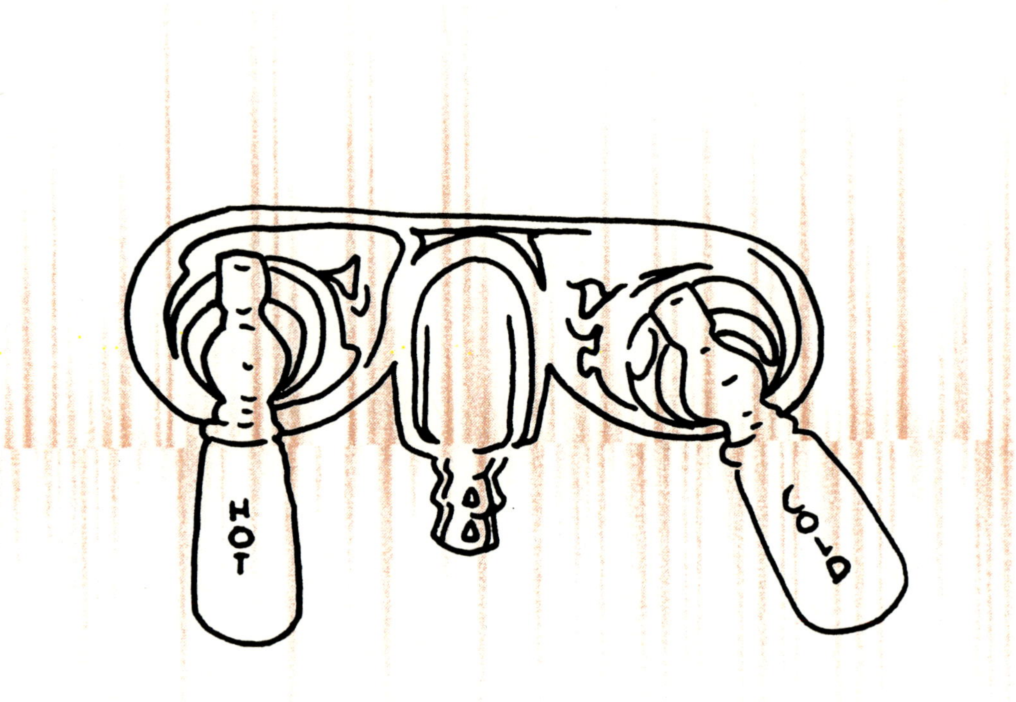

I'm meant to forget that certain very basic gestures—fingers on toggles—are a satisfaction, and to never mind their passing. Am I alone in wanting to turn and adjust faucet handles? Repair fixtures with a wrench and a smear of putty? Yank a towel on a roller to produce a clean spot? Now you need only wave your wet hands at a sensor or move around until lasers kick in. Or often, don't. And then, there you are, swaying dumbly alone in your stall—a scene very different from, say, a man dancing naked, as William

Carlos Williams wrote,

"grotesquely
before my mirror
waving my shirt round my head
and singing softly to myself
'I am lonely, lonely'"[2]

—a loneliness so vital it's worth celebrating, so human it's lyrical, so physically achy it calls forth the drive to *make* something of it.

Or how about this: You've got a drill with a bad battery, a nickel-cadmium, heavy-metal thing you want to recycle. Opening the drill takes a good ten minutes with tiny screwdrivers. The drill itself is just fine: the gears enmeshed and sharp; the motor precise in its cylindrical compactness, nestled within its casing, all the parts good, all able to perform. But the battery pack costs way more than the drill, is specialized and hard to find, or doesn't even exist without the drill, so you throw the whole thing away.

A thing designed with a subsystem that quits, with no way of

2. From William Carlos Williams's "Danse Russe."

keeping the perfect remains. An object lost to itself at the outset. No sense of a tool becoming, through long use, a hand's extension, no hint of its shape responding to a body—of such a fit being intimate.

I'm not supposed to be upset by any of this.

Childhood's a long training in never minding all you're losing, everything that's falling, crashing, being taken. In the diamonded, rhinestoned late 1970s most things were too bright, or tight, or Lucite—and I wasn't learning. I turned away from (and here's a good, bygone word) the *boughten*, and instead toward...well, mostly a dream of wildness, Long Island's marshes long filled in and black-topped for parking lots, malls, Levittowns. I spent much time prospecting for bits. One summer I rose at dawn and gathered dandelions from all the front yards, washed and sugared them, added yeast, water, lemon—and then, for the next few weeks, set to racking off my dandelion wine. (I was reading Bradbury's story at the time.) I wore flannel shirts and overalls, work boots and braids. In love with my bike and my dog, I read outdoor survival guides. I looked for occasions to use the word *pemmican*. I wrote out the steps for tanning hides (the native way, with animal brains, or ash, or acorns). I heard *mall* as the tool—not as a baubly, over-lit place. What others called *wild* in me I knew to be a fending off. A countering. A minding greatly.

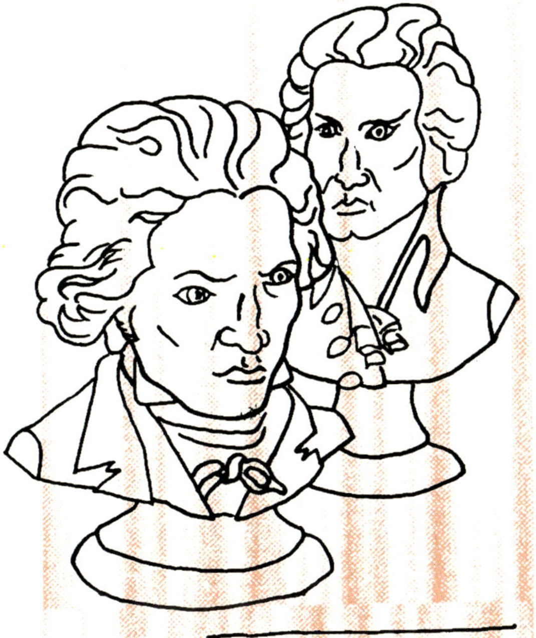

In the Outdoors section of the *Rapid City Journal* I recently read about a couple "Bonded by Their Adventures." Their twenty safaris. Their thirty-year marriage. A few of their tales were gathered therein (in the scary dark of a rainforest when etc. etc. which ended in "the pure joy of being together outdoors")—but it was the front page photo that held me. On the very bright walls, head after head (after head after head)—at least 50 mounts in the living room alone. And how strange the couple looked, alive, among them.

I remember the busts of Beethoven and Mozart (and Haydn and Liszt and Chopin) in my elementary school's music room. I couldn't make any sense of them: a pianist with no arms; a joyless composer who wrote "Ode to Joy." Their limbless bodies in marbly coldness. Stunted and chopped. I knew a head with a bit of neck was meant to be never minded. Another version of how-things-are-done. The men, canonical. The sculptures, memorial. A cliché of sight. I *understood.*

Still it was hard to see anything but severedness.

You should know that the place in *The Scream* where the figure stands is an actual road in Oslo overlooking the bay where screams from the slaughterhouse and asylum converged. Munch's sister was locked in there, while he was free to walk with friends and think and listen and create.

He wrote:

"I went along the road with two friends—
    The sun set
Suddenly the sky became blood—and I felt the breath of sadness
~~A tearing pain beneath my heart~~
I stopped—leaned against the fence—deathly tired

Clouds over the fjord ~~of blood~~ dripped reeking with blood
My friends went on but I just stood trembling with an open wound
in my breast ~~trembling with anxiety~~ I heard a huge extraordinary
scream pass through nature."[3]

You might try it—anyone might: at the end of a day finding yourself abruptly alone, and since certain sounds cannot be unheard, certain images cannot be unseen, in that moment, when a bay might tilt, or a sky drain and pin its red light hard to your chest, you might press your hands to your ears and, at that spot where the world leaks in, wherever it happens—diner, store, street—there, in the moment a scream originates, try to make something of that.

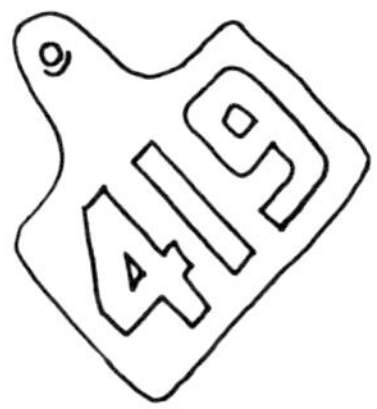

3. From Prideaux's *Edvard Munch: Behind the Scream.*

COLOPHON

The Literary House Press is an imprint
of The Rose O'Neill Literary House
at Washington College in Chestertown, Maryland.
www.washcoll.edu/centers/lithouse

Jehanne Dubrow & James Allen Hall, series editors
Lindsay Lusby, assistant editor
Julie Armstrong, sales representative

This is a trade version of Scream, first printed by hand on Brockway paper
in Futura type by James Dissette on a Vandercook 4 proof press
and bound by Campbell-Logan Bindery.

This essay first appeared in *The Georgia Review* (Fall 2015 Issue),

It was also awarded a Pushcart Prize in 2016.